EXAMPLES

OF THE

Solutions

OF

FUNCTIONAL EQUATIONS.

BY

CHARLES BABBAGE, A.M. F.R.S. L. & E. F.C.P.S.

AND SECRETARY TO THE ASTRONOMICAL SOCIETY OF LONDON.

CAMBRIDGE UNIVERSITY PRESS
Cambridge, New York, Melbourne, Madrid, Cape Town,
Singapore, São Paulo, Delhi, Tokyo, Mexico City

Cambridge University Press
The Edinburgh Building, Cambridge CB2 8RU, UK

Published in the United States of America by
Cambridge University Press, New York

www.cambridge.org
Information on this title: www.cambridge.org/9781107616004

First published 1820
First paperback edition 2013

A catalogue record for this publication is available from the British Library

ISBN 978-1-107-61600-4 Paperback

NOTICE.

THE object of the following Examples of Functional Equations, is to render a subject of considerable interest, more accessible to mathematical students, than it has hitherto been. It is, perhaps, that subject of all others, which most requires the assistance of particular instances, in order fully to comprehend the meaning of its symbols, which are of the most extreme generality; that assistance is also more particularly required in this branch of science, in consequence of its never yet having found its way into an Elementary Treatise.

Oct. 20. 1820.

OF

FUNCTIONAL EQUATIONS.

———◆———

If a function a is of such a form, that, when it is twice performed on a quantity, the result is the quantity itself, or if $a^2(x) = x$, then it is called a periodic function of the second order, if $a^n(x) = x$, then it is termed a periodic function of the nth order, thus when $a(x) = a - x$ the second function, or

$$a(a\,x) = a(a - x) = a - (a - x) = a - a + x = x.$$

If $a(x) = \dfrac{1}{1 - x}$,

then $a^2 x = a(a\,x) = \dfrac{1}{1 - \dfrac{1}{1 - x}} = \dfrac{1 - x}{1 - x - 1} = \dfrac{x - 1}{x}$,

and

$$a^3 x = a^2 a\,x = \frac{a\,x - 1}{a\,x} = \frac{\dfrac{1}{1 - x} - 1}{\dfrac{1}{1 - x}} = 1 - \overline{1 - x} = x,$$

the first of these examples is a periodic function of the second, the last is a periodic function of the third order.

Prob. 1. To find periodic functions of the second order.

Since such functions must satisfy the equation $\psi^2 x = x$, we have

$$\psi x = \psi^{-1} x,$$

or ψ must be such a function, that it shall be the same as its inverse ; if therefore $y = \psi x$, we have also $x = \psi^{-1} y = \psi y$,

† A

or if x and y are connected by some equation, it must be symmetrical relative to x and y; y or ψx must then be determined from the equation

$$* \, F \{ \overline{x, \; \psi x} \} = 0,$$

for instance, if $x + \psi x - a = 0$, $\psi x = a - x$,

or if $x \, \psi \, x = a^2$, $\psi \, x = \dfrac{a^2}{x}$.

Another method of determining such functions is as follows: since ψx is of such a form that $\psi^2 x = x$ any symmetrical function of x and ψx remain constant when x is changed into ψx thus

$$F \{ \overline{x, \; \psi x} \} \text{ becomes } F \{ \overline{\psi x, \; \psi^2 x} \} = F \{ \overline{\psi x, \; x} \},$$

if therefore, we can find any particular solution of the equation $\psi^2 x = x$, containing an arbitrary constant we may substitute such a function for it, but $\psi x = a - x$ is a particular solution therefore

$$\psi x = F(\overline{x, \; \psi x}) - x,$$

or

$$x + \psi x = F(\overline{x, \; \psi x}),$$

and by changing the arbitrary function into another of the same form, we find

$$F \{ \overline{x, \; \psi x} \} = 0,$$

as before.

These two methods of determining periodic functions of the second order, are not so convenient as a third process which can be extended to all orders.

---

* Bars placed above quantities under the functional sign, indicate that the function is symmetrical relative to those quantities.

Assume $\psi x = \phi^{-1} f \phi x$, then

$$\psi^2 x = \phi^{-1} f \phi \phi^{-1} f \phi x = \phi^{-1} f^2 \phi x,$$

this must be equal to x or

$$\phi^{-1} f^2 \phi x = x,$$

this equation will be fulfilled if $f^2 v = v$, or if f is a particular solution, and if also ϕ^{-1} is such an inverse function that $\phi^{-1} \phi v = v$. If therefore ϕ is arbitrary, and f is a particular solution of $f^2 x = x$, then the solution of $\psi^2 x = x$ is

$$\psi x = \phi^{-1} f \phi x.$$

Ex. Let $f x = \dfrac{a}{x}$, then $\psi x = \phi^{-1}\left(\dfrac{a}{\phi x}\right)$,

if $f(x) = \dfrac{a - b x}{b + c x}$, $\psi x = \phi^{-1}\left(\dfrac{a - b \phi x}{b + c \phi x}\right)$;

from these may easily be derived the following periodic functions of the second order,

$$\psi x = a - x \qquad\qquad \psi x = \frac{x}{x - 1}$$

$$\psi x = \frac{x - 2}{x - 1} \qquad\qquad \psi x = \frac{a^2}{x}$$

$$\psi x = \frac{1 - x}{1 + x} \qquad\qquad \psi x = \sqrt{1 - x^2}$$

$$\psi x = \frac{x + 1}{x - 1} \qquad\qquad \psi x = \frac{x}{\sqrt{x^2 - 1}},$$

$$\psi x = \tan^{-1}\left(\frac{\sin (a - x)}{\cos a.\ \cos x}\right) \qquad \psi x = \log (a - \varepsilon^x)$$

$$\psi x = (a^n - x^n)^{\frac{1}{n}} \qquad\qquad \psi x = x - \log (\varepsilon^x - 1)$$

$$\psi x = \frac{x}{(x^n - a^n)^{\frac{1}{n}}} \qquad\qquad \psi x = \tan^{-1}(a - \tan x)$$

PROB. 2.   Required periodic functions of the third order, or such as fulfil the equation $\psi^3 x = x$.

Assume $\psi x = \phi^{-1} f \phi x$, then the equation becomes

$$\psi^3 x = \phi^{-1} f \phi \phi^{-1} f \phi \phi^{-1} f \phi x = \phi^{-1} f^3 \phi x = x,$$

which will be verified if $f(v)$ is a particular solution of $f^3 v = v$, and if ϕ^{-1} is such an inverse value that $\phi^{-1} \phi v = v$, hence the solution of the equation is

$$\psi x = \phi^{-1} f \phi x,$$

one solution is $\dfrac{1}{1-x}$ and hence $\psi x = \phi^{-1} \left(\dfrac{1}{1 - \phi x} \right)$

more particular cases are

$$\psi x = \frac{a^2}{a - x} \qquad\qquad \psi x = \frac{1 + x}{1 - 3x}$$

$$\psi x = \frac{a^2}{ac - c^2 x} \qquad\qquad \psi x = \frac{\sqrt{a x^2 - a^2}}{x}$$

$$\psi x = \frac{a x - a^2}{x} \qquad\qquad \psi x = \frac{1}{1 - x}$$

$$\psi x = \left(\frac{a^2}{a - x^n} \right)^{\frac{1}{n}} \qquad \psi x = -\log(1 - \epsilon^x)$$

$$\psi x = \frac{(a x^n - a^2)^{\frac{1}{n}}}{x} \qquad \psi x = \log(a \epsilon^x - a^2) - x$$

$$\psi x = \frac{a + bx}{c - \dfrac{a^2 + bc + c^2 x}{a}} \qquad \psi x = \log(\epsilon^x - \epsilon^c) - x + c.$$

PROB. 3.   To find periodic functions of the nth order, or to solve the equation $\psi^n x = x$.

Assume as before $\psi x = \phi^{-1} f \phi x$ then it becomes

$$\phi^{-1} f \phi \phi^{-1} f \phi \ldots . \phi^{-1} f \phi x = \phi^{-1} f^n \phi x = x,$$

which is verified if f is a particular solution of $f^n x = x$, and if ϕ^{-1} is such an inverse function that $\phi^{-1}\phi x = x$.

It now remains to find particular solutions of $\psi^n x = x$ which may be accomplished in the following manner : let $f x$ represent $\dfrac{a+b\,x}{c+d\,x}$ then the nth function will be of the same form, or

$$f^n(x) = \frac{A_n + B_n\,x}{C_n + D_n\,x},$$

where A_n, B_n, C_n, D_n, are functions of a, b, c, d, and n, these may be so determined that $D_n = 0$, $A_n = 0$ and $B_n = C_n$ all which conditions are satisfied, if

$$d = -\frac{b^2 - 2\,b\,c\,\cos\dfrac{2\,k\,\pi}{n} + c^2}{\left(2 + 2\cos\dfrac{2\,k\,\pi}{n}\right)a}$$

hence

$$\phi\,x = \phi^{-1}\left\{\frac{a + b\,\phi\,x}{c - \dfrac{b^2 - 2\,b\,c\,\cos\dfrac{2\,k\,\pi}{n} + c^2}{\left(2 + 2\cos\dfrac{2\,k\,\pi}{n}\right)a}\,\phi\,x}\right\}$$

a more detailed account of this method of solution may be found in a paper by Mr. Horner in the Annals of Philosophy, Nov. 1817.

Instances of $\psi^4 x = x$ are

$$\psi\,x = \frac{1}{2}\frac{1}{1-x} \qquad\qquad \psi\,x = \frac{1+x}{1-x}$$

$$\psi\,x = \frac{2}{2-x} \qquad\qquad \psi\,x = \frac{2\,a^2}{2\,a\,c - c^2\,x}$$

$$\psi\,x = 2\,\frac{x-1}{x} \qquad\qquad \psi\,x = \frac{a+b\,x}{c - \dfrac{b^2 + c^2}{2\,a}x}$$

$$\psi x = \frac{\sqrt{2}}{\sqrt{2 - x^2}} \qquad\qquad \psi x = - \sqrt{\frac{1 - x^2}{1 + x^2}}$$

$$\psi x = \frac{(2\, x^n - 2)^{\frac{1}{n}}}{x}$$

$$\psi x = \log 2 - x + \log (\varepsilon^x - 1).$$

All those cases which satisfy the equation $\psi^2 x = x$, also fulfil that of $\psi^4 x = x$, as well as all those which fulfil any of these equations $\psi^2 x = - x$, $\psi^2 x = \dfrac{1}{x}$, or more generally $\psi^2 x = \alpha\, x$, where $\alpha\, x$ is a particular solution of the equation $\psi^2 x = x$.

The following particular cases satisfy the equation $\psi^6 x = x$.

$$\psi x = \frac{1}{3(1 - x)} \qquad\qquad \psi x = \frac{3x - 1}{3x}$$

$$\psi x = \frac{3}{3 - x} \qquad\qquad \psi x = \frac{3 a^2}{3 ac - c^2 x}.$$

$$\psi x = 3\,\frac{x - 1}{x} \qquad\qquad \psi x = \frac{3 + 3x}{3 - x}$$

$$\psi x = \frac{a + bx}{c - \dfrac{b^2 - bc + c^2}{3a} x}$$

$$\psi x = \frac{1}{x}\left(x^n - \frac{1}{3}\right)^{\frac{1}{n}}$$

$$\psi x = \log 3 - x + \log (\varepsilon^x - 1).$$

The principle on which the solution of the functional equation $F\{\, x,\ \psi x,\ \psi \alpha x\,\} = 0$ depends, where $\alpha^2 x = x$, is that by substituting αx for x we have another equation $F\{\, \alpha x,\ \psi \alpha x,\ \psi x\,\} = 0$, between which and the given equation we may eliminate $\psi \alpha x$ and the result will be the value of ψx a few examples will illustrate this method.

(1).   Given $\psi(x) + a\psi(-x) = x^n$

by putting $-x$ for x this becomes

$$\psi(-x) + a\psi(x) = (-x)^n,$$

and eliminating $\psi(-x)$, we have

$$\psi x - a^2 \psi x = x^n - a(-x)^n,$$

hence

$$\psi x = \frac{1 - (-1)^n a}{1 - a^2} x^n.$$

(2).   Given $\psi x - a\psi \dfrac{1}{x} = \varepsilon^x$

put $\dfrac{1}{x}$ for x, $\psi \dfrac{1}{x} - a\psi x = \varepsilon^{\frac{1}{x}}$

and

$$\psi x - a\,\varepsilon^{\frac{1}{x}} - a^2 \psi x = \varepsilon^x,$$

$$\psi x = \frac{\varepsilon^x + a\,\varepsilon^{\frac{1}{x}}}{1 - a^2}$$

(3).   Given $(\psi x)^2 . \psi \dfrac{1-x}{1+x} = c^3 x$

put $\dfrac{1-x}{1+x}$ for x, it becomes

$$\left(\psi \frac{1-x}{1+x}\right)^2 . \psi x = c^3 \frac{1-x}{1+x},$$

eliminating $\psi \dfrac{1-x}{1+x}$ by means of the former, we find

$$\psi x = \left(\frac{1+x}{1-x} c^3 x^2\right)^{\frac{1}{3}}.$$

(4). Given $\psi x + \dfrac{1}{1-x^2} \psi \sqrt{(1-x^2)} = 1 + x^2$

putting $\sqrt{(1-x^2)}$ for x, we have

$$\psi \sqrt{(1-x^2)} + \frac{1}{x^2} \psi x = 2 - x^2,$$

and substituting this value of $\psi \sqrt{1-x^2}$ in the former equation

$$\psi x + \frac{2-x^2}{1-x^2} - \frac{1}{x^2-x^4} \psi x = 1 + x^2,$$

hence

$$\left(\frac{x^2-x^4-1}{x^2-x^4}\right) \psi x = 1 + x^2 - \frac{2-x^2}{1-x^2} = \frac{-1+x^2-x^4}{1-x^2}$$

and $\psi x = x^2$.

(5). Given $\dfrac{\psi x}{1+\psi x} + x \dfrac{\psi(-x)}{1+\psi(-x)} = 1$

put $\psi_1 x = \dfrac{\psi x}{1+\psi x}$ thus the equation becomes

$\psi_1 x + x \psi_1(-x) = 1$, and changing x into $-x$ we have $\psi_1(-x) - x \psi_1(x) = 1$, by which eliminating $\psi(-x)$ from the former, we find

$$\psi_1 x = \frac{1-x}{1+x^2},$$

hence

$$\psi x = \frac{\psi_1 x}{1-\psi_1 x} = \frac{1-x}{x+x^2}.$$

(6). Given $\psi x + \dfrac{1+x}{x} \psi \dfrac{1}{x} = c,$

putting $\dfrac{1}{x}$ for x this becomes

$$\psi \frac{1}{x} + (1 + x)\, \psi\, x = x$$

and by eliminating $\psi \frac{1}{x}$, we have

$$\psi\, x = \frac{1}{1 + x + x^2}\, c$$

(7). Given $\psi\, x + x\, \psi\, (1 - x) = 1,$
putting $1 - x$ for x, we have

$$\psi\, (1 - x) + (1 - x)\, \psi\, (x) = 1,$$

whence, by elimination,

$$\psi\, x = \frac{1 - x}{1 - x(1 - x)} = \frac{1 - x}{1 - x + x^2}.$$

(8). Given $\dfrac{\psi\, x}{\psi\, x - x} + x\, \dfrac{\psi\, (1 - x)}{\psi\, (1 - x) + x - 1} = 1,$

put $\psi_1\, x = \dfrac{\psi\, x}{\psi\, x - x}$, then will $\psi_1(1 - x) = \dfrac{\psi\, (1 - x)}{\psi\, (1 - x) + x} \; 1$,
and the equation becomes

$$\psi_1\, x + x\, \psi_1\, (1 - x) = 1,$$

the same as in the last example; let $f\, x$ represent the solution there found, then

$$\psi_1\, x = f\, x = \frac{\psi\, x}{\psi\, x - x},$$

whence

$$\psi\, x = \frac{x f\, x}{f\, x - 1},$$

if we take for $f\, x$ its value $\dfrac{1 - x}{1 - x + x^2}$, we have

$$\psi\, x = \frac{x - 1}{x}.$$

In case the equation is symmetrical with regard to ψx and $\psi \alpha x$, the process of elimination apparently becomes illusory. By a peculiar artifice this difficulty may be overcome, and it happens rather singularly that in all these cases, the solution which is so obtained contains an arbitrary function, and in general the solution is the most extensive which the question admits of.

(9). Given $\psi x = \psi \dfrac{1}{x}$.

If we put $\dfrac{1}{x}$ for x, this is changed into $\psi \dfrac{1}{x} = \psi x$, the same as the given equation ; it is therefore impossible to eliminate.

Let us now suppose $\psi x = a \psi \dfrac{1}{x} + b,$

which becomes the given equation when $a = 1$ and $b = 0$. By putting $\dfrac{1}{x}$ for x this is changed into

$$\psi \frac{1}{x} = a \psi x + b,$$

and eliminating $\psi \dfrac{1}{x}$, we have

$$\psi x = \frac{ab + b}{1 - a^2} = \frac{b}{1 - a},$$

if $b = 0$ and $a = 1$, this becomes a vanishing fraction whose value is any constant quantity c, and we have $\psi x = c$, which fulfils the equation. This is a very limited solution, but the following plan will lead us to much more general ones.

Take the equation

$$\psi x = a \psi \frac{1}{x} + v \phi x,$$

which coincides with the given one when $v = 0$ and $a = 1$; also ϕx is any arbitrary function of x ; putting $\dfrac{1}{x}$ for x, we have

$$(11)$$

$$\psi\frac{1}{x} = a\,\psi x + v\,\phi\frac{1}{x},$$

and by elimination,

$$\psi x = \frac{a\,\phi\frac{1}{x} + \phi x}{1 - a^2}\,v.$$

Let a become $1 + 0$ and v become 0 at the same time, then

$$\frac{0}{1 - (1 + 2.0 + 0^2)} = \frac{0}{-2.0 + 0^2} = \frac{1}{-2 + 0} = -\frac{1}{2}$$

and the solution becomes

$$\psi x = -\frac{\phi\frac{1}{x} + \phi x}{2},$$

or changing the arbitrary function

$$\psi x = \phi x + \phi\frac{1}{x},$$

in which ϕ is indefinite.

This solution is, in fact, nothing more than an arbitrary symmetrical function of x and $\frac{1}{x}$, and may be expressed thus

$$\psi x = \chi\left(\overline{x}, \frac{1}{x}\right).$$

Precisely the same course of reasoning will produce the solutions of the following equation.

(10). $\psi(x) = \psi(a - x)$

$$\psi x = \chi(\overline{x}, \overline{a - x}).$$

(11). $\psi(x) = \psi\left(\dfrac{1 - x}{1 + x}\right),$ $\psi x = \chi\left\{\overline{x}, \overline{\dfrac{1 - x}{1 + x}}\right\}.$

(12)

(12). $\quad \psi(x) = \psi\left(\dfrac{x}{\sqrt{(x^2-1)}}\right), \quad \psi\,x = \chi\left\{\bar{x}, \,\overline{\dfrac{x}{\sqrt{(x^2-1)}}}\right\}.$

(13). $\quad \psi\left(\dfrac{x}{2-x}\right) = \psi(1-x), \quad \psi\,x = \chi\left\{x, \,\overline{\dfrac{1-x}{1+x}}\right\}.$

(14). $\quad \psi\,x = \psi(\alpha\,x), \quad \psi\,x = \chi(x, \,\alpha\,x),$ where $\alpha^2\,x = x.$

(15). The objection which has just been stated occurs in the equation $\psi\,x + \psi\left(\dfrac{x}{x-1}\right) = c,$

and a similar mode of proceeding will obviate it. The given equation is a particular case of

$$\psi\,x + a\,\psi\left(\frac{x}{x-1}\right) = c + v\,\phi\,x,$$

with which it coincides, if $a=1$ and $v=0$; putting $\dfrac{x}{x-1}$ for x in this, we have

$$\psi\left(\frac{x}{x-1}\right) + a\,\psi\,x = c + v\,\phi\left(\frac{x}{x-1}\right),$$

and elimination produces

$$\psi\,x = \frac{c-ac}{1-a^2} + \left\{\phi\,x - a\,\phi\left(\frac{x}{x-1}\right)\right\}\frac{v}{1-a^2}$$

If $v=0$ and $a=1$, this gives

$$\psi\,x = \frac{c}{2} + \phi\,x - \phi\left(\frac{x}{x-1}\right)$$

in which the function ϕ has been changed into another similar one.

16. Given $\psi(1+a) + \psi(1-x) = 1 - x^2,$

put $x-1$ for x, then

(13)

$$\psi x + \psi(2-x) = 1 - (x-1)^2 = 2x - x^2,$$

this is a particular case of the equation

$$\psi x + a\psi(2-x) = 2x - x^2 + v\phi x,$$

with which it agrees if $v=0$ and $a=1$; changing x into $2-x$ and eliminating $\psi(2-x)$ from the result, we find

$$\psi x = \frac{2x - x^2 - a(2x - x^2)}{1 - a^2} + \{\phi x - a\phi(2-x)\}\frac{v}{1-a^2}.$$

or

$$\psi x = \frac{2x - x^2}{1+a} + \{\phi x - a\phi(2-x)\}\frac{v}{1-a^2},$$

If $a=1$ and $v=0$, we have

$$\psi x = \frac{2x - x^2}{2} + \phi x - \phi(2-x).$$

(17.) Given $\dfrac{1}{x + \psi x} + \dfrac{x}{1 + x\psi\frac{1}{x}} = 2,$

put $\dfrac{1}{x + \psi x} = \psi_1 x$ then $\dfrac{1}{\frac{1}{x} + \psi\frac{1}{x}} = \psi_1\frac{1}{x} = \dfrac{x}{1 + x\psi\frac{1}{x}},$

and the equation becomes

$$\psi_1 x + \psi_1\frac{1}{x} = 2,$$

whose solution may be found by the method just explained to be

$$\psi_1 x = 1 + \phi x - \phi\frac{1}{x},$$

hence

$$\psi x = \frac{1}{1 + \phi x - \phi\frac{1}{x}} - x.$$

(18). Required the equation of that class of curves which possess the following property, (Part IV. Fig. 1.) a given abscissa $AB = a$ being taken, then the product of any two ordinates at equal distances from B, shall always be equal to the square of the abscissa a. If $y = \psi x$ represent the equation of the curve, then the condition expressed analytically is

$$\psi(a - x) \cdot \psi(a + x) = a^2.$$

Putting $a - x$ for x, and then $\log \psi(x) = \psi_1 x$, we have

$$\psi_1 x + \psi_1(2a - x) = 2 \log a,$$

whose solution is

$$\psi_1 x = \log a + \phi x - \phi(2a - x),$$

hence

$$\log \psi x = \log a + \phi x - \phi(2a - x),$$

and

$$\psi x = \epsilon^{\log a} \times \epsilon^{\phi x} \times \epsilon^{-\phi(2a-x)} = \frac{a \, \epsilon^{\phi x}}{\epsilon^{\phi(2a - x)}}$$

and changing the arbitrary function ϕ into $\log \phi$,

$$\psi x = a \frac{\phi x}{\phi(2a - x)},$$

and the class of curves are comprehended in the equation

$$y = \frac{a \, \phi x}{\phi(2a - x)}.$$

(19). Given the equation

$$\{ \psi x \}^2 + \left\{ \psi \left(\frac{\pi}{2} - x \right) \right\}^2 = 1.$$

This is the equation on which the composition of forces is made to depend in the Mecanique Cœleste, p. 5.

Put $\psi_1 x$ for $(\psi x)^2$ then it becomes

$$\psi x + \psi_1\left(\frac{\pi}{2} - x\right) = 1.$$

which is a particular case of

$$\psi_1 x + a \psi_1\left(\frac{\pi}{2} - x\right) = 1 + v \phi x.$$

Substituting $\frac{\pi}{2} - x$ for x, this gives

$$\psi_1\left(\frac{\pi}{2} - x\right) + \psi_1 x = 1 + v \phi\left(\frac{\pi}{2} - x\right),$$

and eliminating $\psi_1\left(\frac{\pi}{2} - x\right)$, we have

$$\psi x = \frac{1}{1 + a} - \frac{\phi x - a \phi\left(\frac{\pi}{2} - x\right) v}{1 - a^2}.$$

making $a = 1$ and $v = 0$, and changing ϕx in $2\phi x$, we have

$$\psi_1 x = \frac{1}{2} - \phi x + \phi\left(\frac{\pi}{2} - x\right)$$

and therefore

$$\psi x = \sqrt{\frac{1}{2} - \phi x + \phi\left(\frac{\pi}{2} - x\right)}$$

In case the coefficient of $\psi a x$ in the equation $\psi x + f x \psi a x = f x$, is of such a form that $f x \cdot f a x = 1$, the denominator will vanish, and we must then have recourse to an artifice similar to that which has already been explained.

(20).   **Ex. 1.** Let $\psi x + x^{2n} \psi \frac{1}{x} = x^n$.

put $\psi x + (x^{2n} + v \phi x) \psi \frac{1}{x} = x^n$

which coincides with the given equation if $v=0$; then changing x into $\frac{1}{x}$, we have

$$\psi\frac{1}{x} + \left(x^{-2n} + v\,\phi\,\frac{1}{x}\right)\psi\,x = x^{-n}$$

and by elimination

$$\psi\,x = \frac{x^n - (x^{2n} + v\,\phi\,r)x^{-n}}{1 - (x^{2n} + v\,\phi\,x)\left(x^{-2n} + v\,\phi\,\frac{1}{x}\right)}$$

$$= \frac{x^{-n}\,\phi\,x}{x^{2n}\,\phi\,\frac{1}{x} + x^{-2n}\,\phi\,x + v\,\phi\,x\,.\,\phi\,\frac{1}{x}}$$

and when v vanishes,

$$\psi\,x = \frac{x^{-n}\,\phi\,x}{x^{2n}\,\phi\,\frac{1}{x} + x^{-2n}\,\phi\,x}, \qquad (a).$$

The equation in this example may be solved differently, as follows. Multiply by x^{-n} and it becomes

$$x^{-n}\,\psi\,x + x^n\,\psi\,\frac{1}{x} = 1,$$

put $\psi_1\,x = x^{-n}\,\psi\,x$, then

$$\psi_1\,x + \psi_1\frac{1}{x} = 1,$$

whose general solution found by a process already explained is

$$\psi_1\,x = \frac{1}{2} - \phi\,x + \phi\,\frac{1}{x},$$

hence

$$\psi\,x = \frac{x^n}{2} - x^n\,\phi\,x + x^n\,\phi\,\frac{1}{x}.$$

This solution differs in form from that which was previously found, but it may be proved to be the same by the following substitution ; since ϕ is quite an arbitrary function

$$\phi x = x^{2n} \left(1 - 2 \phi_1 x + 2 \phi_1 \frac{1}{x} \right),$$

this gives $x^{2n} \phi \frac{1}{x} + x^{-2n} \phi x = 2$ and (a) becomes

$$\psi x = \frac{x^n}{2} - x^n \phi_1 x + x^n \phi_1 \frac{1}{x},$$

exactly as the last solution.

(21). **Given**

$$\left(\frac{\psi x + 1}{\psi x - 1} \right)^n - \left(\frac{\psi \left(\frac{x}{x-1} \right) + 1}{\psi \left(\frac{x}{x-1} \right) - 1} \right)^n = \frac{(x-1)^2 - 1}{x - 1};$$

put $\psi_1 x = \frac{\psi x + 1}{\psi x - 1}$, then it becomes

$$\psi_1 x - \psi_1 \left(\frac{x}{x-1} \right) = \frac{(x-1)^2 - 1}{x - 1},$$

which is a particular case of

$$\psi_1 x + (a + v \phi x) \psi_1 \left(\frac{x}{x-1} \right) = \frac{(x-1)^2 - 1}{x - 1}$$

with which it agrees, when $a = -1$ and $v = 0$.

Put$\frac{x}{x-1}$ for x, then

$$\psi_1 \frac{x}{x-1} + \left(a + v \phi \frac{x}{x-1} \right) \psi_1 x = - \frac{(x-1)^2 - 1}{x - 1},$$

whence by elimination,

† c

$$\psi_1 x = \frac{(x-1)^2 - 1}{x-1} \cdot \frac{1 + a + v\phi x}{1 - (a + v\phi x)\left(a + v\phi\, \dfrac{x}{x-1}\right)},$$

and when $a = -1$, and $v = 0$, this becomes

$$\psi_1 x = \frac{x^2 - 2x}{x-1} \cdot \frac{\phi x}{\phi x + \phi\left(\dfrac{x}{x-1}\right)},$$

and restoring the value of $\psi_1 x$, we find

$$\psi x = \frac{\left\{\dfrac{x^2 - 2x}{x-1} \cdot \dfrac{\phi x}{\phi x + \phi\left(\dfrac{x}{x-1}\right)}\right\}^{\frac{1}{n}} + 1}{\left\{\dfrac{x^2 - 2x}{x-1} \cdot \dfrac{\phi x}{\phi x + \phi\left(\dfrac{x}{x-1}\right)}\right\}^{\frac{1}{n}} - 1}$$

(22). $\psi x + \dfrac{x}{\sqrt{(1+x^2)}} \quad \psi \sqrt{(1-x^2)} = x$;

$$\psi x = \frac{x(1-x^2)\phi x}{x^2 \phi \sqrt{(1-x^2)} + (1-x^2)\phi x} :$$

this solution was found by pursuing the course so frequently pointed out : another but not a more general one may be obtained as follows : multiply by $\sqrt{(1-x^2)}$; then

$$\sqrt{(1-x^2)}\,\psi x + x\,\psi\,\sqrt{(1-x^2)} = x\,\sqrt{(1-x^2)},$$

putting $\sqrt{(1-x^2)}\,\psi x = \psi_1 x$, we have

$$\psi_1 x + \psi_1 \sqrt{(1-x)} = x\sqrt{(1-x^2)},$$

whose solution is $\psi_1 x = \dfrac{x\sqrt{(1-x^2)}}{2} + \phi x - \phi\sqrt{(1-x^2)}$,

hence

$$\psi x = \frac{x}{2} + \frac{\phi x - \phi\sqrt{(1-x^2)}}{\sqrt{(1-x^2)}}.$$

It would not be difficult to shew the identity of these two apparently different solutions.

(23). Given the equation

$$\frac{\psi\, x}{\sqrt{[\psi\,(1-x)]}} + \frac{\psi\,(1-x)}{\sqrt{(\psi\, x)}} = 1,$$

put $\psi_1\, x = \dfrac{\psi\, x}{\sqrt{[\psi\,(1-x)]}}$, then $\psi_1\,(1-x) = \dfrac{\psi\,(1-x)}{\sqrt{(\psi\, x)}}$;

and the equation becomes

$$\psi_1\, x + \psi_1\,(1-x) = 1,$$

whose general solution is

$$\psi_1\, x = \frac{\phi\, x}{\phi\, x + \phi\,(1-x)};$$

hence

$$\frac{(\psi\, x)^2}{\psi\,(1-x)} = \frac{(\phi\, x)^2}{[\phi\, x + \phi\,(1-x)]^2}$$

putting $1 - x$ for x, and eliminating $\psi\,(1-x)$, we find

$$\psi\, x = \frac{\{ (\phi\, x)^2\, \phi\,(1-x) \}^{\frac{2}{3}}}{[\phi\,(1-x) + \phi\, x]^2}.$$

(24). Given $(\psi\, x)^2 + \left(\psi\, \dfrac{a^2}{x} \right)^2 = \dfrac{x^4 + a^4}{x^2}\, \psi\, x\, .\, \psi\, \dfrac{a^2}{x}$;

divide by $\psi\, x\, .\, \psi\, \dfrac{a^2}{x}$ then

$$\frac{\psi\, x}{\psi\, \dfrac{a^2}{x}} + \frac{\psi\, \dfrac{a^2}{x}}{\psi\, x} = \frac{x^4 + a^4}{x^2};$$

putting $\psi_1\, x = \dfrac{x}{\psi\, \dfrac{a^2}{x}}$, this becomes

$$\psi_1 x + \psi_1 \frac{a^2}{x} = \frac{x^4 + a^4}{x^2},$$

a particular solution of which is $\psi_1\, x = x^2$; hence

$$\frac{\psi\, x}{\psi\, \frac{a^2}{x}} = x^2 \text{ or } \psi\, x = x^2 \psi\, \frac{a^2}{x}$$

and the general solution of this is

$$\psi\, x = x\, \chi\left(\overline{x,\ \frac{a^2}{x}}\right).$$

(25).   Given $\dfrac{\psi\, x + x}{\psi\, \dfrac{1}{x}} + x\, \dfrac{1 + x\, \psi\, \dfrac{1}{x}}{\psi\, x} = 1 + x^3,$

put $\psi_1\, x = \dfrac{\psi\, x + x}{\psi\, \dfrac{1}{x}}$, then the equation becomes

$$\psi_1\, x + x^2\, \psi_1\, \frac{1}{x} = 1 + x^3,$$

whose solution is $\psi_1\, x = \dfrac{(x^2 + 1)\, \phi\, x}{\phi\, x + x^4\, \phi\, \dfrac{1}{x}};$

hence $\dfrac{x + \psi\, x}{\psi\, \dfrac{1}{x}} = \dfrac{(x^2 + 1)\, \phi\, x}{\phi\, x + x^4 \phi\, \dfrac{1}{x}};$

putting $\dfrac{1}{x}$ for x and eliminating $\psi\, \dfrac{1}{x}$, we find

$$\psi\, x = \frac{(2\,x + x^{-1})\, \phi\, x + x^5\, \phi\, \dfrac{1}{x}}{(x^2 + x^{-2})\, \phi\, x \cdot \phi\, \dfrac{1}{x} - x^{-4}(\phi\, x)^2 - x^{-4}\left(\phi\, \dfrac{1}{x}\right)^2}\left(\phi\, \frac{1}{x} + x^{-4}\phi\, x\right)$$

(26). Given $(\psi x)^m . (\psi - x)^n - (\psi x)^n . (\psi - x)^m = 2x$; putting $\psi_1 x = (\psi x)^m . (\psi - x)^n$, it becomes

$$\psi_1 x - \psi_1 (- x) = 2x,$$

whose solution is $\psi_1 x = x + \chi (\overline{x, \; - x})$,

hence

$$(\psi x)^m . (\psi - x)^n . = x + \chi (\overline{x, \; - x}),$$

and by the process for eliminating $\psi (- x)$, we shall find

$$\psi x = \left\{ \frac{\chi (\overline{x, \; - x}) + x}{\{ \chi (\overline{x, \; - x}) - x \}^{\frac{m}{n}}} \right\}^{\frac{m}{m^2 - n^2}}.$$

(27). Given $\psi x + fx . \psi \alpha x = f_1 x$, where αx is such a function of x that $\alpha^2 x = x$; putting αx for x, we have

$$\psi \alpha x + f \alpha x . \psi x = f_1 \alpha x,$$

and by eliminating $\psi \alpha x$

$$\psi x = \frac{f_1 x - fx . f_1 \alpha x}{1 - fx . f \alpha x}.$$

If $fx . f\alpha x = 1$ and $f_1 x - fx . f_1 \alpha x = 0$, then the solution becomes a vanishing fraction; also the general value of $f_1 x$ is in that case $f_1 x = \sqrt{(fx)} . f_2 (\overline{x, \; \alpha x})$ and the equation becomes

$$\psi x + fx . \psi \alpha x = \sqrt{fx} . f_2 (\overline{x, \; \alpha x});$$

dividing this by $\sqrt{(fx)}$ and putting instead of $\dfrac{1}{\sqrt{fx}}$ its value $\sqrt{(f\alpha x)}$ derived from the equation $fx . f\alpha x = 1$, we have

$$\sqrt{(f\alpha x)} . \psi x + \sqrt{(fx)} . \psi \alpha x = f_2 (\overline{x, \; \alpha x}),$$

which is a symmetrical equation, whose general solution is

$$\sqrt{(f\alpha x)} . \psi x = \frac{f_2 (\overline{x, \; \alpha x}) . \phi x}{\phi x + \phi \alpha x},$$

hence

$$\downarrow x = \frac{\sqrt{(f x)} \cdot f_2(\overline{x, \ \overline{a\,x}}) \cdot \phi\,x}{\phi\,x + \phi\,a\,x} .$$

(28). Given $\quad a + b \downarrow x = \downarrow (a + b\,x),$

$$\downarrow x = a\,\frac{1 - b^n}{1 - b} + b^n x.$$

(29). Given $\quad \dfrac{a \downarrow x}{b + c \downarrow x} = \downarrow \left(\dfrac{a\,x}{b + c\,x}\right);$

$$\downarrow x = \frac{a^n x}{b^n + c\,\dfrac{a^n - b^n}{a - b}\,x} .$$

(30). Given $\quad \dfrac{\downarrow x}{\downarrow x - 1} = \downarrow \left(\dfrac{x}{x - 1}\right);$

$$\downarrow x = \frac{x}{\dfrac{1 - (- 1)^n}{2}\,x + (- 1)^n} ,$$

where n is arbitrary; it may therefore be changed into any symmetrical function of x and $\dfrac{x}{x - 1}$.

(31). The three last examples are particular cases of the equation

$$a \downarrow x = \downarrow a\,x,$$

whose general solution is $\downarrow x = a^n x.$

(32). Given $\quad \dfrac{\downarrow \left(\dfrac{x}{1 + x}\right)}{1 + \downarrow \left(\dfrac{x}{1 + x}\right)} = \downarrow \left(\dfrac{x}{1 + 2\,x}\right),$

$$\downarrow x = \frac{x}{1 + n\,x} .$$

(33). Given $a \downarrow a\,x = \downarrow a^2\,x$

$$\downarrow x = a^n\,x.$$

(34). Given $\downarrow x + a \downarrow \left(\dfrac{1}{1-x}\right) = \dfrac{1}{x}$;

$\dfrac{1}{1-x}$ is a periodic function of the third order, or of the

form $a^3\,x = x$; putting $\dfrac{1}{1-x}$ for x, we have

$$\downarrow \left(\frac{1}{1-x}\right) + a \downarrow \left(\frac{x-1}{x}\right) = 1 - x,$$

and in this again putting $\dfrac{1}{1-x}$ for x, we find

$$\downarrow \left(\frac{x-1}{x}\right) + a \downarrow x = \frac{x}{x-1} ;$$

$\downarrow \left(\dfrac{1}{x-1}\right)$ and $\downarrow \left(\dfrac{x-1}{x}\right)$ being eliminated between these

three equations, we have

$$\downarrow x = \frac{1}{1+a^3} \left\{ \frac{1}{x} - a(1-x) + a^2 \frac{x}{x-1} \right\}.$$

(35). Given $\downarrow x - a \downarrow \left(\dfrac{\sqrt{(x^2-1)}}{x}\right) = x^{2n}$,

put $\dfrac{\sqrt{(x^2-1)}}{x}$ for x, it becomes

$$\downarrow \left(\frac{\sqrt{(x^2-1)}}{x}\right) - a \downarrow \left(\frac{1}{\sqrt{(1-x^2)}}\right) = \frac{(x^2-1)^n}{x^{2n}}.$$

Again, put $\dfrac{\sqrt{(x^2-1)}}{x}$ for x, and we have

$$\downarrow \left(\frac{1}{\sqrt{(1-x^2)}}\right) - a \downarrow x = \frac{1}{(1-x^2)^n},$$

by eliminating $\psi\left(\dfrac{\sqrt{(x^2-1)}}{x}\right)$ and $\psi\left(\dfrac{1}{\sqrt{(1-x^2)}}\right)$ from these three equations, we shall find

$$\psi x = \frac{1}{1-a^3}\left\{x^{3n} + a\,\frac{(x^2-1)^n}{x^{3n}} + a^2\,\frac{1}{(1-x^2)^n}\right\}.$$

(36). Given $\psi x + \psi\left(\dfrac{1+x}{1-3x}\right) + \psi\left(\dfrac{x-1}{3x+1}\right) = a$

the function $\dfrac{1+x}{1-3x}$ is periodic of the third order, and by the process of elimination

$$\psi x = \frac{a\,\phi x + \phi_1 x - \phi_1\left(\dfrac{1+x}{1-3x}\right)}{\phi x + \phi\left(\dfrac{1+x}{1-3x}\right) + \phi\left(\dfrac{x-1}{3x+1}\right)}.$$

(37). Given $\dfrac{1}{\sqrt{\psi x}} + \dfrac{1}{\sqrt{\psi\dfrac{x-1}{x}}} + \dfrac{1}{\sqrt{\psi\dfrac{1}{1-x}}} = a.$

Putting $\psi_1 x = \dfrac{1}{\sqrt{\psi x}}$, we have

$$\psi_1 x + \psi_1\frac{x-1}{x} + \psi_1\frac{1}{1-x} = a,$$

whose solution is

$$\psi_1 x = \frac{a\,\phi x + \phi_1 x - \phi_1\dfrac{1}{1-x}}{\phi x + \phi\dfrac{1}{1-x} + \phi\dfrac{x-1}{x}},$$

hence

$$\psi x = \left\{\frac{\phi x + \phi\dfrac{1}{1-x} + \phi\dfrac{x-1}{x}}{a\,\phi x + \phi_1 x - \phi_1\dfrac{1}{1-x}}\right\}^2.$$

(38). Given $\psi x \cdot \psi \dfrac{1}{1-x} \cdot \psi \dfrac{x-1}{x} = c^3.$

Putting $\psi, x = \log \psi x$, we find

$$\psi_1 x + \psi_1 \frac{1}{1-x} + \psi_1 \frac{x-1}{x} = \log (c^3),$$

whose solution is found in the last problem. Changing ϕ_1 into $\log \phi_1$, we have

$$\psi x = \frac{\phi_1 x}{\phi_1 \dfrac{1}{1-x}} \log^{-1} \left(\frac{3\,\phi x \cdot \log c}{\phi x + \phi \dfrac{x-1}{x} + \phi \dfrac{1}{1-x}} \right)$$

Similarly if $a\,x$ be any periodic equation of the third order.

(39). $\quad \psi x + \psi a x + \psi a^2 x = a,$

has for its solution

$$\psi x = \frac{a\,\phi x + \phi_1 x - \phi_1 a x}{\phi x + \phi a x + \phi a^3 x}.$$

(40). $\quad \psi x \cdot \psi a x \cdot \psi a^2 x = c^3,$

has for its solution

$$\psi x = \frac{\phi_1 x}{\phi a x} \log^{-1} \left\{ \frac{3\,\phi x \cdot \log c}{\phi x + \phi a x + \phi a^3 x} \right\}.$$

(41). Given $\psi x + f x \cdot \psi a x = f_1 x$, where $a^3 x = x$,

Putting successively $a x$ and $a^2 x$ for x, we have

$$\psi a x + f a x \cdot \psi a^2 x = f_1 a x,$$

$$\psi a^2 x + f a^2 x \, \psi x = f_1 a^2 x,$$

and eliminating $\psi a x$ and $\psi a^2 x$ from these three equations, we have

$$\psi x = \frac{f_1 x - f x \cdot f_1 a x + f x \cdot f a x \cdot f_1 a^2 x}{1 + f x \cdot f a x \cdot f a^2 x}.$$

† D

(42). Given $\psi x + f x . \psi a x = f_1 x$ where $a^n x = x,$ a similar process of elimination will produce

$$\psi x = \frac{f_1 x - f x . f_1 a x + \ldots f x . f a x . f a^{n-2} x . f_1 a^{n-1} x}{1 - (-1)^n f x . f a x .. f a^{n-1} x}.$$

(43). Given the equation

$$1 + f x . (\psi x + \psi a x) - \psi x . \psi a x = 0,$$

where $a x$ is a periodic function of the second order, and $f x$ is any function symmetrical relative to x and $a x$

$$f x = \frac{\psi x . \psi a x - 1}{\psi x + \psi a x},$$

consider ψx and $\psi a x$ as two variables, and differentiate with respect to them, then

$$\frac{d \psi x}{1 + (\psi x)^2} + \frac{d \psi a x}{(1 + \psi a x)^2} = 0,$$

and by integration,

$$\tan^{-1} \psi x + \tan^{-1} \psi a x = C = - \frac{1}{f x},$$

whose complete solution is

$$\tan^{-1} \psi x = \frac{\phi x}{\phi x + \phi a x} \tan^{-1} \frac{-1}{f x},$$

hence

$$\psi x = \tan \left\{ \frac{\phi x}{\phi x + \phi a x} \tan^{-1} \frac{-1}{f x} \right\},$$

this process is analogous to one employed by M. Laplace, for the integration of a similar equation of differences.

---

* *Journal de l'Ecole Polytecnique*, Cah. 15.

(44). Given

$$\psi x + \psi \left(\frac{1+x}{1-x}\right) + \psi \left(-\frac{1}{x}\right) + \psi \left(\frac{x-1}{x+1}\right) = 1.$$

$\dfrac{1+x}{1-x}$ being a periodic function of the 4th order.

$$\psi x = \cfrac{\phi x}{\phi x + \phi\left(\frac{1+x}{1-x}\right) + \phi\left(\frac{-1}{x}\right) + \phi\left(\frac{x-1}{x+1}\right)}$$

$$+ \frac{x^2+1}{x-1} \chi \left(\overline{x,\ \frac{1+x}{1-x},\ \frac{-1}{x},\ \frac{x-1}{x+1}}\right) +$$

$$+ \frac{x^2+1}{x.(1-x)} \chi \left\{\overline{x,\ \frac{1+x}{1-x},\ \frac{-1}{x},\ \frac{x-1}{x+1}}\right\}.$$

(45). Given $\psi (x, y) + \psi \left(\dfrac{a^2}{x}, \dfrac{a^2}{y}\right) = 1,$

$$\psi (x, y) = \frac{\phi (x, y)}{\phi (x, y) + \phi \left(\dfrac{a^2}{x}, \dfrac{a^2}{y}\right)}.$$

(46). $\psi (x, y) + \psi \left(\dfrac{a^2}{x}, -y\right) = y^2,$

$$\psi (x\ y) = \frac{y^2 \phi (x, y)}{\phi (x, y) + \phi\left(\dfrac{a^2}{x}, -y\right)}.$$

(47). Given $\psi (x, y) + x^2 \psi \left(\dfrac{a^2}{x}, \dfrac{y}{y-1}\right) = \dfrac{x y^2}{y-1},$

$$\psi (x, y) = \frac{(y-1)^{-1} x y^2 \phi (x, y)}{\phi (x, y) + \phi \left(x, \dfrac{y}{y-1}\right)}.$$

(48). Given $\psi (x, y) + f (x, y) \psi (a x, \beta y) =$

$$\sqrt{f(x, y)}.f_1(\overline{x,\ a x},\ \overline{y,\ \beta y}),$$

where $a^2 x = x$, $\beta^2 y = y$ and $f(x, y)$ is such a function that

$$f(x, y).f(ax, \beta y) = 1,$$

then $\psi(x, y) = \dfrac{\sqrt{f(ax, \beta y)}.f_1(\bar{x}, \,\overline{ax}, \,\bar{y}^{\frac{1}{}}, \,\overline{\beta y}^{\frac{1}{}})}{f(x, y)\phi(ax, \beta y) + f(ax, \beta y)\phi(x, y)},$$

(49). Given $\psi(x, y) = \psi\left(\dfrac{x-y}{y}, \dfrac{x-y}{x}\right)$

$$\psi(x, y) = \phi\left(\dfrac{x}{y}\right).$$

(50). Given $\psi(x, y) = \psi\left(\dfrac{f(x, y)}{y}, \dfrac{f(x, y)}{x}\right),$

$$\psi(x, y) = \phi\left(\dfrac{x}{y}\right).$$

(51). Given $\psi(x, y) = \psi\left(\dfrac{y}{2}\sqrt{\dfrac{y}{2x}}, \sqrt{2xy}\right),$

$$\psi(x, y) = \chi\left(2xy + y^2, \dfrac{1}{y\sqrt{2xy}}\right).$$

(52). Given $\psi(x, y) = \psi(y, x)$
$$\psi(x, y) = \chi(x + y, xy).$$

(53). Given $\psi(x, y) = \left(\dfrac{x}{y}\right)^3 \psi(y, x)$

$$\psi(x, y) = \dfrac{x}{y^2}.\phi(\bar{x}, \bar{y}) = \dfrac{x}{y^2}.\phi_1(x + y, xy).$$

(54). Given $\psi(\pi - x) = \dfrac{d\,\psi x}{dx},$

differentiating $\dfrac{d}{dx}\psi(\pi - x) = \dfrac{d^2\,\psi x}{dx^2},$

putting $\pi - x$ for x in the given equation

$$\psi (x) = - \frac{d \psi (\pi - x)}{d x},$$

and eliminating $\dfrac{d \psi (\pi - x)}{d x}$, we have

$$- \psi x = \frac{d^2 \psi x}{d x^2},$$

whence by integration

$$\psi x = b \cos x + c \sin x,$$

and it will be found that $c = - b$; hence

$$\psi x = b (\cos x - \sin x).$$

(55). Given $\psi (x, y) = \dfrac{d \psi (x, a - y)}{d x}$,

put $a - y$ for y, then

$$\psi (x, a - y) = \frac{d \psi (x, y)}{d x},$$

differentiate this relative to x, then

$$\frac{d \psi (x, a - y)}{d x} = \frac{d^2 \psi (x, y)}{d x^2},$$

which being substituted in the given equation produces

$$\psi (x, y) = \frac{d^2 \psi (x, y)}{d x^2},$$

whose solution is

$$\psi (x, y) = \epsilon^x \phi y + \epsilon^{-x} \phi, y,$$

ϕ and ϕ_1 being two arbitrary functions so constituted as to fulfil the given equation, in order to determine them, put $a - y$ for y and differentiate relative to x, then

$$\frac{d \psi (x, a - y)}{d x} = \epsilon^x \phi (a - y) - \epsilon^- \phi_1 (a - y)$$

hence

$$\phi y = \phi (a - y) \text{ and } \phi_1 y = - \phi_1 (a - y),$$

whose solutions are

$$\phi x = \chi(\overline{y, \ a - y}) \text{ and } \phi_1 y = (a - 2y) \chi_1 (\overline{y, \ a - y}),$$

hence the general solution of the equation is

$$\psi (x, \ y) = \varepsilon^x \chi (\overline{y, \ a-y}) + \varepsilon^{-x} (a - 2y) \chi_1 (\overline{y, \ a - y})$$

A similar mode of solution is applicable to the three follow-ing equations.

(56). Given $\psi (x, \ y) = \dfrac{d}{dx} \psi \left(x, \ \dfrac{1}{y} \right)$

$$\psi (x, y) = \varepsilon^x \phi \left(\overline{y, \ \dfrac{1}{y}} \right) + \varepsilon^{-x} \dfrac{1-y^2}{y} \phi_1 \left(\overline{y, \ \dfrac{1}{y}} \right).$$

(57). Given $\psi (x, \ y) = \dfrac{d}{dx} \psi \left(x, \ \dfrac{y}{y-1} \right)$

$$\psi (x, \ y) = \varepsilon^x \phi \left(\overline{y, \ \dfrac{y}{y-1}} \right) + \varepsilon^{-x} \dfrac{2y - y^2}{y-1} \phi_1 \left(\overline{y, \ \dfrac{y}{y-1}} \right).$$

(58). Given $\psi (x, \ y) = \dfrac{d}{dx} \psi (x, \ a y)$, where $a^2 y = y$

$$\psi (x, \ y) = \varepsilon^x \phi (y, \ a y) + \varepsilon^{-x} (a y - y) \phi_1 (\overline{y, \ a y}).$$

(59). Given $\psi (x, \ y) = \dfrac{d \psi (x, \ a y)}{dx}$

where a is such a function that $a^4 y = y$.

Substituting successively $a y$, $a^2 y$, $a^3 y$ for y, we have

$$\psi (x, \ a y) = \dfrac{d \psi (x, \ a^2 y)}{dx},$$

$$\psi(x, \ a^2 y) = \frac{d\,\psi(x, \ a^3 y)}{d\,x},$$

$$\psi(x, \ a^3 y) = \frac{d\,\psi(x, \ y)}{d\,x}.$$

From the given equation $\dfrac{d\,\psi(x, \ a y)}{d\,x}$ may be eliminated by means of the second, and from the result $\dfrac{d\,\psi(x, \ a^2 y)}{d\,x}$ may be eliminated by the third equation, and continuing this, we should find

$$\psi(x, \ y) = \frac{d^4\,\psi(x, \ y)}{d\,x^4},$$

the solution of this equation is

$$\psi(x, \ y) = \epsilon^x\,\phi\,y + \epsilon^{-x}\,\phi_1\,y + \sin x \cdot \phi_2 y + \cos x \cdot \phi_3\,y,$$

$\phi, \ \phi_1, \ \phi_2, \ \phi_3$ must be determined so as to satisfy the given equation, taking the differential and putting $a\,y$ for y, we have

$$\frac{d\,\psi(x, \ a y)}{d\,x} = \epsilon^x\,\phi\,a\,y - \epsilon^{-x}\,\phi_1\,a\,y + \cos x \cdot \phi_2\,a\,y - \sin x \cdot \phi_3\,a\,y$$

the first condition to satisfy is

$$\phi\,y = \phi\,a\,y,$$

which gives

$$\phi\,y = \chi\,(\overline{y}, \ \overline{a\,y}, \ \overline{a^2\,y}, \ \overline{a^3\,y}),$$

the next condition is

$$\phi_1\,y = -\ \phi_1\,a\,y$$

whose solution is

$$\phi_1\,y = (a^3\,y - a^2 y + a\,y - y\,(\chi_1\,(\overline{y}, \ \overline{a\,y}, \ \overline{a^2 y}, \ \overline{a^3\,y}),$$

the other two conditions are

$$\phi_2 y = -\ \phi_3\,a\,y, \text{ and } \phi_3\,y = \phi_2\,a\,y$$

putting αy for y in the second of these it becomes $\phi_3\,\alpha y = \phi_2\,\alpha^2\,y$ and this substituted in the first gives,

$$\phi_2 y = -\ \phi_2\,\alpha^2\,y,$$

whose solution is $\phi_2 y = (\alpha^2 y - y)\,\chi_2(\overline{y},\ \overline{\alpha y}^{\frac{1}{}},\ \overline{\alpha^2 y},\ \overline{\alpha^3 y}^{\frac{1}{}})$

hence

$$\phi_3 y = (\alpha^3 y - \alpha y)\,\chi_2(\overline{\alpha y},\ \overline{\alpha^2 y},\ \overline{\alpha^3 y}^{\frac{1}{}},\ \overline{y}^{\frac{1}{}}),$$

and the general solution of the equation is

$$\psi(x,\ y) = \epsilon^x\,\chi(\overline{y},\ \overline{\alpha y},\ \overline{\alpha^2 y},\ \overline{\alpha^3 y}) +$$

$$+\ \epsilon^{-x}(\alpha^3 y - \alpha^2 y + \alpha y - y)\,\chi_1(\overline{y},\ \overline{\alpha y},\ \overline{\alpha^2 y},\ \overline{\alpha^3 y}) +$$

$$+\ (\alpha^2 y - y)\,\chi_2(\overline{y},\ \overline{\alpha y}^{\frac{1}{}},\ \overline{\alpha^2 y},\ \overline{\alpha^3 y}^{\frac{1}{}})\sin x +$$

$$+\ (\alpha^3 y - \alpha y)\,\chi_2(\overline{\alpha y},\ \overline{\alpha^2 y}^{\frac{1}{}},\ \overline{\alpha^3 y},\ \overline{y}^{\frac{1}{}})\cos x.$$

(60). Given the equation $\psi(x,\ y) = \dfrac{d^n\,\psi(x,\ \alpha y)}{d\,x^n}$,

where α is such a function that $\alpha^n x = x$.

This equation may be reduced to the solution of the partial differential equation

$$\psi(x,\ y) = \frac{d^{pn}\,\psi(x,\ y)}{d\,x^{pn}},$$

and the arbitrary functions of y which occur in its solution, must be determined by the conditions of the equation.

(61). Given the equation

$$\frac{d\,\psi(a - x,\ y)}{d\,y} = \frac{d\,\psi(x,\ b - y)}{d\,x},$$

put $a - x$ for x, also $b - y$ for y, then we have the two equations

(33)

$$\frac{d\psi(x,\,y)}{dy} = -\frac{d\psi(a-x,\,b-y)}{dx},$$

$$-\frac{d\psi(a-x,\,b-y)}{dy} = \frac{d\psi(x,\,y)}{dx}.$$

If the first of these be differentiated relative to y, and the second relative to x; then the right side of the first resulting equation will be identical with the left side of the second, and we shall have

$$\frac{d^2\psi(x,\,y)}{dy^2} = \frac{d^2\psi(x,\,y)}{dx^2};$$

the solution of this partial differential equation is

$$\psi(x,\,y) = \phi(x+y) + \phi_1(x-y);$$

the two arbitrary functions must be determined so as to satisfy the equation; we have

$$\frac{d\psi(a-x,\,y)}{dy} = \phi'(a-x+y) - \phi'_1(a-x-y),$$

$$\frac{d\psi(x,\,b-y)}{dx} = \phi'(b+x-y) + \phi'_1(-b+x+y),$$

ϕ' and ϕ'_1 being the differential coefficients of ϕ and ϕ_1 these two expressions must be identical, hence

$$\phi'(a-\overline{x-y}) = \phi'(b+\overline{x-y}),$$

and

$$-\phi'_1(a-\overline{x+y}) = \phi'_1(-b+\overline{x+y}),$$

the solutions of which equations are

$$\phi'(x+y) = \chi\{\overline{x+y},\ \overline{a-b-x-y}\},$$

and

$$\phi'_1(x+y) = (a-b-2x-2y)\chi_1\{\overline{x+y},\ \overline{a-b-x-y}\}$$

and substituting these values, we have

† E

(34)

$$\psi(x, y) = \int(dx + dy)\,\chi\,\{\,\overline{x + y,\ a - b - x - y}\,\} +$$
$$+ \int(dx - dy)(a - b - 2x - 2y)\,\chi_1\,\{\,\overline{x - y,\ a - b - 2x + 2y}\,\}.$$

(63). Given the equation

$$\frac{d\psi\left(x, \dfrac{1}{y}\right)}{dx} = \frac{d\psi\left(\dfrac{1}{x}, y\right)}{dy}.$$

Put $\dfrac{1}{y}$ for y and differentiate relative to x, then

$$\frac{d^2\psi(x, y)}{dx^2} = -\frac{d^2\psi\left(\dfrac{1}{x}, \dfrac{1}{y}\right)}{dx\,dy}\,y^2.$$

Again, put $\dfrac{1}{x}$ for x, and differentiate relative to y, then

$$-\frac{d^2\psi\left(\dfrac{1}{x}, \dfrac{1}{y}\right)}{dx\,dy}\,x^2 = \frac{d^2\psi(x, y)}{dy^2}$$

hence

$$\frac{d^2\psi(x, y)}{dy^2} = \frac{x^2}{y^2}\frac{d^2\psi(x, y)}{dx^2},$$

the solution of this equation of partial differentials is

$$\psi(x, y) = x\,\phi\left(\frac{x}{y}\right) + \phi_1(xy):$$

to determine the form of ϕ and ϕ_1, we have

$$\frac{d\psi\left(x, \dfrac{1}{y}\right)}{dx} = \phi(xy) + xy\,\phi'(xy) + \frac{1}{y}\,\phi'_1\left(\frac{x}{y}\right),$$

$$\frac{d\psi\left(\dfrac{1}{x}, y\right)}{dy} = -\frac{1}{x^2 y}\,\phi'\left(\frac{1}{xy}\right) + \frac{1}{x}\,\phi'_1\left(\frac{y}{x}\right).$$

In order that these two expressions may coincide, we must have

$$\phi(xy) + xy\,\phi'(xy) = -\frac{1}{x^2 y^2}\phi'\left(\frac{1}{xy}\right)$$

$$\phi'_1\left(\frac{x}{y}\right) = \frac{y}{x}\phi'_1\left(\frac{y}{x}\right).$$

The first of these multiplied by $d(xy)$ may be put under the form

$$d(xy).\,\phi(xy) + xy\,\frac{d\phi(xy)}{d(xy)}d(xy) = d\phi\left(\frac{1}{xy}\right)$$

whose integral is

$$xy\,\phi(xy) = \phi\left(\frac{1}{xy}\right),$$

the solution of which functional equation is

$$\phi(xy) = \frac{1}{\sqrt{xy}}\chi\left(\overline{xy},\ \overline{\frac{1}{xy}}\right)$$

the solution of the second equation is

$$\phi'_1\left(\frac{x}{y}\right) = \sqrt{\frac{y}{x}}\cdot\chi_1\left(\overline{\frac{x}{y}},\ \overline{\frac{y}{x}}\right),$$

employing these values of ϕ and ϕ_1, we have

$$\psi(x,\,y) = \sqrt{xy}\cdot\chi\left(\overline{\frac{x}{y}},\ \overline{\frac{y}{x}}\right) +$$

$$\int d(xy).\left(\frac{1}{xy}\right)^{\frac{1}{2}}\chi_1\left(\overline{xy},\ \overline{\frac{1}{xy}}\right).$$

(63). Given $\dfrac{d\psi(x,\,\alpha y)}{dx} = \dfrac{d\psi(\beta x,\,y)}{dy}$,

where $\alpha^2 y = y$ and $\beta^2 x = x$ a process nearly similar to that

(36)

by which the two last equations were solved will lead to the partial differential equation

$$\frac{d\,a\,y}{dy}\cdot\frac{d^2\,\psi\,(x,\,y)}{d\,x^2}=\frac{d\,\beta\,x}{d\,x}\cdot\frac{d^2\,\psi\,(x,\,y)}{d\,y^2}.$$

(64). Given $\quad\psi\,a\,x=\psi\,\psi\,x=\psi^2 x.$

It is evident, that whatever be the form of a, this equation can always be satisfied by assuming $\psi\,x=a\,x$, hence the solutions of the following equations,

$$\psi\,(-x)=\psi^2\,x \qquad\qquad \psi\,x=-x$$

$$\psi\left(\frac{a\,x}{b\,+\,c\,x}\right)=\psi^2 x \qquad \psi\,x=\frac{a\,x}{b\,+\,c\,x}$$

$$\psi\sqrt{\frac{1\,+\,x}{x}}=\psi^2\,x \qquad \psi\,x=\sqrt{\frac{1\,+\,x}{x}},$$

(65). Given $\quad\psi\,(2a-x)=\psi^3\,x.$

Put $\quad\psi\,x=\phi^{-1}f\phi\,x,$

then $\quad\psi^2\,x=\phi^{-1}f\phi\phi^{-1}f\phi\,x=\phi^{-1}f^2\phi\,x,$

and $\quad\psi^3\,x=\phi^{-1}f^2\phi\phi^{-1}f\phi\,x=\phi^{-1}f^3\phi\,x,$

and the equation becomes

$$\phi^{-1}f\phi\,(2\,a-x)=\phi^{-1}f^3\,\phi\,x.$$

This equation may be satisfied in the following manner: by making f a periodic function of the second order, we have $f^2v=v$, and the equation becomes

$$\phi^{-1}f\phi\,(2\,a-x)=\phi^{-1}f\phi\,x,$$

or

$$\phi\,(2\,a-x)=\phi\,x,$$

This is satisfied by making ϕ any symmetrical function of x and $2\,a - x$. As an example take $f v = -\,v$, also

$$\phi\, x = x \,.\, \overline{2\,a - x} = 2\,a\,x - x^2,$$

then

$$\phi^{-1}\, x = a \pm \sqrt{(a^2 - x)},$$

and

$$\psi\, x = \phi^{-1} f \phi\, x = a \pm \sqrt{a^2 - \frac{1}{x^2 - 2\,a\,x}},$$

(66). Given $\psi \left(\dfrac{1 - x}{1 + x} \right) = \psi^3\, x$

$\psi\, x = \phi^- f \phi\, x$ where ϕ and f are determined by the equations

$$\phi\, x = \chi \left\{ \overline{\frac{1 - x}{1 + x}}, \; \overline{x} \right\} \quad \text{and} \quad f^2\, x = x.$$

(67). Given $\psi\, a\, x = \psi^3\, x$, where $a^n v = v$,

putting $\psi\, x = \phi^{-1} f \phi\, x$, we have

$$\phi^{-1} f \phi\, a\, x = \phi^{-1} f^3\, \phi\, x$$

determine ϕ from the condition $\phi\, x = \phi\, a\, x$; hence,

$$\phi\, x = \chi \left\{ \overline{x}, \; \overline{a\, x}, \ldots \ldots \overline{a^{n-1}\, x} \right\},$$

and let f be such a function that $f^2\, x = x$, then the equation is satisfied.

(68). Given $\psi^p\, a\, x = \psi^q\, x$ where $q > p$ and $a^n x = x$, the substitution $\phi^{-1} f \phi\, x$ instead of ψ will give,

$$\phi^{-1} f^p\, \phi\, a\, x = \phi^{-1} f^q\, \phi\, x,$$

and this is satisfied if $\phi\, x = \chi \, (\overline{x}, \; \overline{a\, x}, \; \overline{a^{n-1}\, x})$

and also $f^{q-p}\, v = v$, for it then becomes

$$\phi^{-1} f^p\, \phi\, a\, x = \phi^{-1} f^p\, \phi\, x, \quad \text{where } \phi\, x = \phi\, a\, x.$$

(38)

If in a function of two variables, as $\psi\,(x,\,y)$, we substitute the function itself instead of one of those quantities, the result is denoted thus,

$$\psi\,\{\,x,\;\psi\,(x,\,y)\,\} \;=\; \psi^{1,\,2}\,(x,\,y),$$

$$\psi\,\{\,\psi\,(x,\,y),\;y\,\} \;=\; \psi^{2,\,1}\,(x,\,y),$$

if the function itself is substituted simultaneously for x and y, it is denoted thus

$$\psi\,\{\,\psi\,(x,\,y),\;\psi\,(x,\,y)\,\} \;=\; \psi^{\overline{2,\,2}}\,(x,\,y).$$

(69). Given $\quad \psi^{\overline{2,2}}\,(x,\,y) = a,$

By means of the substitution $\phi^{-1}f\phi\,x$, for $\psi\,x$, we are enabled to reduce functional equations of any order to those of the first, a substitution nearly resembling it, will be of equal value for those which contain two or more variables, by assuming

$$\psi\,(x,\,y) \;=\; \phi^{-1}f(\phi\,x,\;\phi\,y),$$

we have

$$\psi^{\overline{2,2}}(x,\,y) = \phi^{-1}f\{\,\phi\phi^{-1}f(\phi\,x,\;\phi\,y),\;\phi\phi^{-1}f(\phi\,x,\;\phi\,y)\,\}$$

$$= \phi^{-1}f^{\overline{2,2}}(\phi\,x,\;\phi\,y),$$

and substituting this value in the equation

$$\phi^{-1}f^{\overline{2,2}}(\phi\,x,\;\phi\,y) = a.$$

Put $\phi^{-1}x$ for x, and ϕ^{-1} for y, also taking the function ϕ on both sides

$$f^{\overline{2,2}}(x,\,y) = \phi\,a.$$

If therefore we are acquainted with a particular solution, we find the general one ; let the function $A\dfrac{x}{y}$ be tried, then

$$f^{\overline{2,2}}(x, y) = A\, \dfrac{A\dfrac{x}{y}}{A\dfrac{x}{y}} = \phi\, a$$

hence $A = \phi\, a$, and the solution is

$$\psi(x, y) = \phi^{-1}\left(\frac{\phi\, x}{\phi\, y}\, \phi\, a\right),$$

a variety of solutions may be found of different forms, such as

$$\psi(x, y) = \frac{a}{\phi(1)}\phi\left(\frac{x}{y}\right), \quad \psi(x, y) = \frac{a}{\phi(1)}\phi\left(\frac{\alpha(x, y)}{\beta(x, y)}\right),$$

where α and β are any two homogeneous functions of the same degree.

(70). If $\psi(x, y) = a\,x + b\,y$,

then $\psi^{\overline{n, n}}(x, y) = (a + b)^{n-1}(a x + b y)$,

(71). If $\psi(x, y)$ is any homogeneous function of x, and y of the degree n,

then

$$\psi^{\overline{k, k}}(x, y) = \{\psi(x, y)\}^{\frac{k-1}{n}} \times \{\psi(1, 1)\}^{\frac{1-n^{k-1}}{1-n}}$$

(72). Given $\psi^{\overline{2, 2}}(x, y) = \sqrt{\psi(x, y)}$,

$$\psi(x, y) = \sqrt{\frac{x^2 + y^2}{2\, y\, \phi\left(\frac{x}{y}\right)}}\, \phi(1).$$

(73). Given

$$\psi^{\overline{2, 2}}(x, y) = \psi(x, y) + \frac{1}{\psi(x, y)},$$

(40)

$$\psi(x,\,y) = \frac{2\,\phi\left(\frac{x}{y}\right)}{(x+y)\,\phi(1)} + \frac{(x+y)\,\phi(1)}{2\,\phi\left(\frac{x}{y}\right)}.$$

(74). Given $\psi^{\overline{2,2}}(x,\,y) = \dfrac{1 - \psi(x,\,y)}{1 + \psi(x,\,y)}$,

$$\psi(x,\,y) = \frac{y\,\phi(1) - x^2\,\phi\left(\frac{x}{y}\right)}{y\,\phi(1) + x^2\,\phi\left(\frac{x}{y}\right)}.$$

(75). $\psi^{\overline{2,2}}(x,\,y) = F\psi(x,\,y)$,

$$\psi(x,\,y) = F\left(\frac{\alpha(x,\,y)}{\beta(x,\,y)}\right),$$

provided α and β are homogeneous with respect to x and y; the first of the $n+1$ degree, the second of the nth, and also at the same time $\alpha(1,\,1) = \beta(1,\,1)$.

(76). Given $\psi^{\overline{2,2}}(x,\,y) = F\psi(x,\,y)$.

Another solution of the same equation is

$$\psi(x,\,y) = F\left(\frac{\alpha(x,\,y)}{\beta(x,\,y)}x\right),$$

where α and β are two such functions, that when $x = y$, we have also

$$\alpha(x,\,y) = \beta(x,\,y).$$

(77). Given $\psi^{\overline{3,3}}(x,\,y) = \psi(x,\,y)$,

$$\psi(x,\,y) = \left\{a - \left(\frac{x^2+y^2}{x+y}\right)^n\right\}^{\frac{1}{n}}.$$

(78). **Given** $\psi^{\overline{n,n}}(x, y) = \{\psi(x, y)\}^m$

$$\psi(x, y) = \left\{ \frac{2xy\, \phi(1)}{(x+y)\, \phi\left(\dfrac{x}{y}\right)} \right\}^{m^{\frac{1}{n-1}}}.$$

(79.) **Given** $\psi^{\overline{4,4}}(x, y) = \{\psi^{\overline{2,2}}(x, y)\}^2$

$$\psi(x, y) = \left\{ \frac{(x+y)\, \phi(1)}{2\, \phi\left(\dfrac{x}{y}\right)} \right\}^{\sqrt{2}}.$$

(80). **Given** $\psi^{\overline{3,3}}(x, y) = \dfrac{x^2}{y}$.

$$\psi(x, y) = \phi^{-1} \left\{ \frac{1 + \phi\, \dfrac{x^2}{y}}{1 - 3\, \phi\, \dfrac{x^2}{y}} \right\}.$$

(81). **Given** $x\,\psi^{1,2}(x, y) = y\,\psi^{2,1}(x, y)$,

put $\phi^{-1} f(\phi x, \phi y)$ for ψx, then it becomes

$$x\, \phi^{-1} f^{1,2}(\phi x, \phi y) = y\, \phi^{-1} f^{2,1}(\phi x, \phi y);$$

putting $\phi^{-1} x$ for x, and $\phi^{-1} y$ instead of y, we have

$$\phi^{-1} x \cdot \phi^{-1} f^{1,2}(x, y) = \phi^{-1} y \cdot \phi^{-1} f^{2,1}(x, y);$$

if $f^{1,2}(x, y) = y$, and $f^{2,1}(x, y) = x$, this equation, becomes identical; but making $f(x, y) = a - x - y$, these two equations are verified; consequently the general solution is

$$\psi(x, y) = \phi^{-1}(a - \phi x - \phi y).$$

(82). **Given** $\psi^{1,2}(x, y) \cdot \psi^{2,1}(x, y) = xy$,

$$\psi(x, y) = \phi^{-1}\left(\frac{a}{\phi x \cdot \phi y}\right).$$

(83). **Given** $x\,\psi^{\overline{2,2}}(x, y) = a\,\psi^{2,1}(x, y)$,

$$\psi(x, y) = \phi^{-1}\left(\frac{\phi a \cdot \phi y}{\phi x}\right).$$

Various methods for the solution of Functional Equations may be found in the following writings :

Speculationes Analytico Geometricæ, *N. Fuss.* Mem. de l'Acad. Imp. de St. Petersburg, Vol. IV. p. 225. 1811.

Memoirs of the Analytical Society, p. 96. 1813.

Observations on various points of Analysis, Phil. Trans. *J. F. W. Herschel.*

Essay towards the Calculus of Functions, *C. Babbage.* 1815.

Ditto, Part II. p. 179. 1816.

Observations on the analogy which subsists between the Calculus of Functions, and other branches of Analysis, Phil. Trans. 1817. p. 197. *C. Babbage.*

Spence's Essays, 1819. Note by *J. F. W. Herschel,* p. 151.

Annals of Philosophy, Nov. 1817. *Mr. Horner.*

Journal of the Royal Institution. *C. Babbage.*

www.ingramcontent.com/pod-product-compliance
Ingram Content Group UK Ltd.
Pitfield, Milton Keynes, MK11 3LW, UK
UKHW010851090126
466816UK00011B/161